BODY SY...

The Skeletal System

by Kay Manolis

Consultant:
Molly Martin, M.D.
Internal Medicine
MeritCare, Bemidji, MN

BELLWETHER MEDIA • MINNEAPOLIS, MN

Note to Librarians, Teachers, and Parents:

Blastoff! Readers are carefully developed by literacy experts and combine standards-based content with developmentally appropriate text.

Level 1 provides the most support through repetition of high-frequency words, light text, predictable sentence patterns, and strong visual support.

Level 2 offers early readers a bit more challenge through varied simple sentences, increased text load, and less repetition of high-frequency words.

Level 3 advances early-fluent readers toward fluency through increased text and concept load, less reliance on visuals, longer sentences, and more literary language.

Level 4 builds reading stamina by providing more text per page, increased use of punctuation, greater variation in sentence patterns, and increasingly challenging vocabulary.

Level 5 encourages children to move from "learning to read" to "reading to learn" by providing even more text, varied writing styles, and less familiar topics.

Whichever book is right for your reader, Blastoff! Readers are the perfect books to build confidence and encourage a love of reading that will last a lifetime!

This edition first published in 2016 by Bellwether Media, Inc.

No part of this publication may be reproduced in whole or in part without written permission of the publisher. For information regarding permission, write to Bellwether Media, Inc., Attention: Permissions Department, 6012 Blue Circle Dr., Minnetonka, MN 55343.

Library of Congress Cataloging-in-Publication Data
Manolis, Kay.
  Skeletal system / by Kay Manolis.
      p. cm. – (Blastoff! readers: body systems)
  Includes bibliographical references and index.
  Summary: "Introductory text explains the functions and physical concepts of the skeletal system with color photography and simple scientific diagrams. Intended for students in grades three through six"–Provided by publisher.
  ISBN: 978-1-60014-247-5 (hardcover : alk. paper)
  ISBN: 978-1-62617-474-0 (paperback : alk. paper)
  1. Human skeleton–Juvenile literature. I. Title.
QM101.M325 2009
612.7'5–dc22   2008032703

Text copyright © 2009 by Bellwether Media, Inc. BLASTOFF! READERS and associated logos are trademarks and/or registered trademarks of Bellwether Media, Inc.
Printed in the United States of America, North Mankato, MN.

Contents

What Is the Skeletal System?	4
Joints	8
How Bones Help You Move	10
Bones Protect Organs	12
What's Inside Bones?	18
Glossary	22
To Learn More	23
Index	24

What Is the Skeletal System?

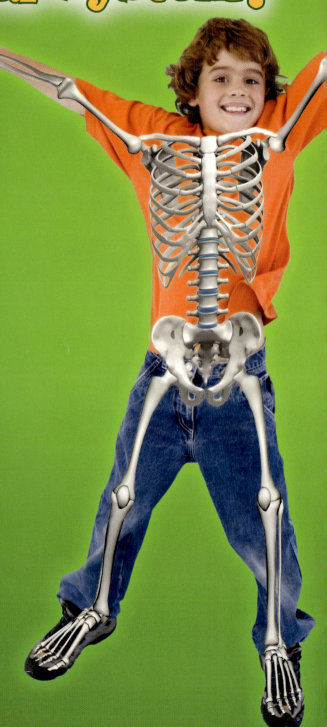

Move your bones! When you walk, run, or dance, your skeletal system is helping you move. The bones in your skeletal system also give your body its shape and protect many of your **organs**.

fun fact

Babies are born with more than 300 bones. Over time, bones grow in size and many grow together to form the adult skeleton, which has 206 bones.

Adult Skeleton

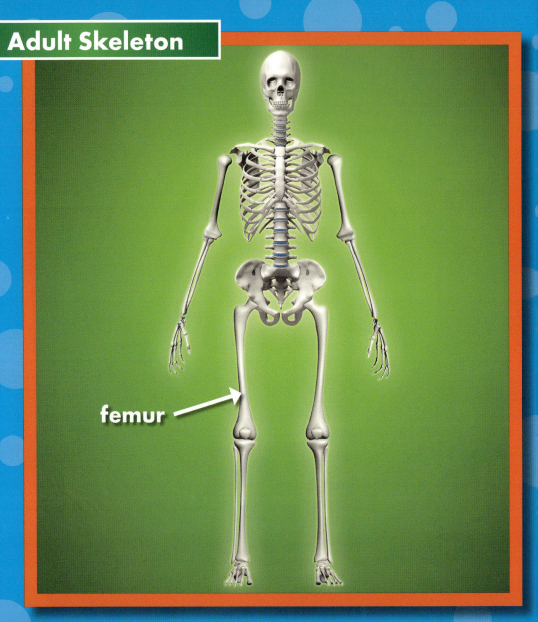

An adult human skeleton has 206 bones. Bones come in many different shapes and sizes. The femur, in the upper leg, is the largest and heaviest bone in the body. The average adult femur is 19 inches (48 centimeters) long.

Hand Bones

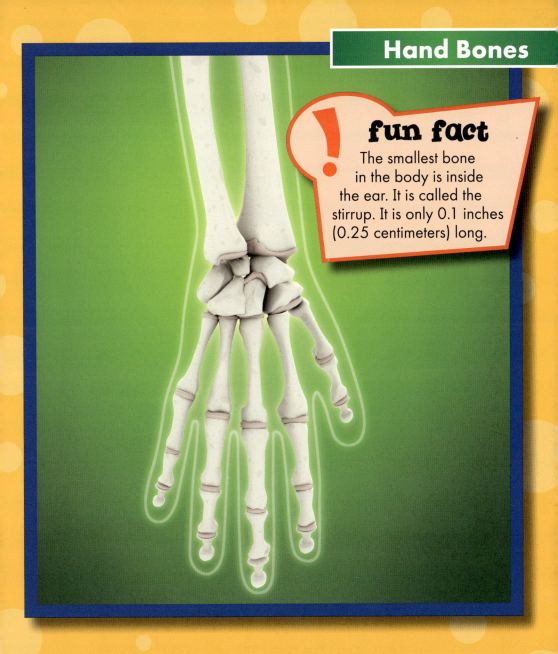

fun fact
The smallest bone in the body is inside the ear. It is called the stirrup. It is only 0.1 inches (0.25 centimeters) long.

A human has 27 bones in each hand! There are three bones in each finger. Thumbs have two bones.

Joints

Knee Joint

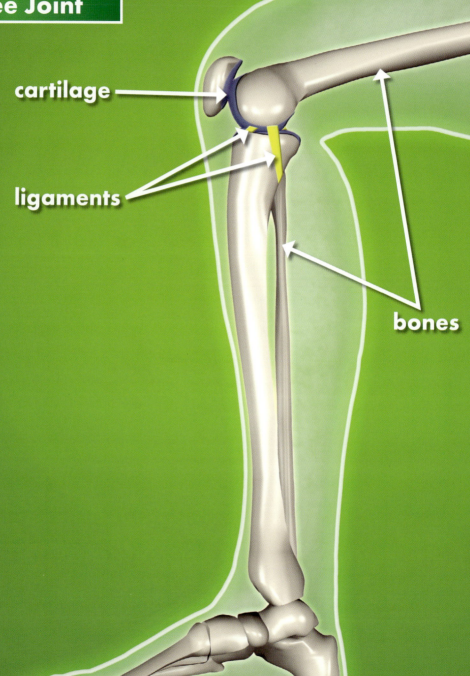

Joints are places where bones come together. Bands of **ligaments** or **cartilage** hold bones together in joints.

Most joints are places where your skeleton can bend or move. For example, hinge joints allow body parts to bend back and forth. The knee joint is your body's biggest hinge joint. Your knee joint bends whenever you walk, run, or sit.

How Bones Help You Move

Bones and joints cannot move by themselves. They depend on other systems of your body.

Before bones can move, **nerves** carry messages from your brain. The messages tell your **muscles** how to move. Muscles pull on bones to move your skeleton.

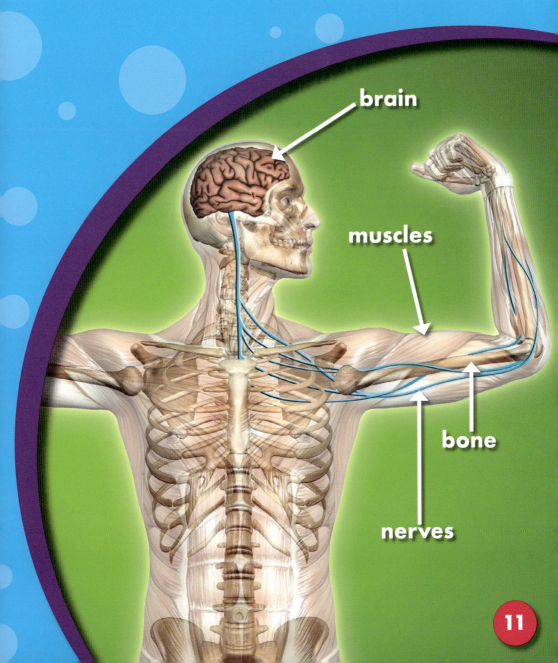

Bones Protect Organs

Bones as Protection

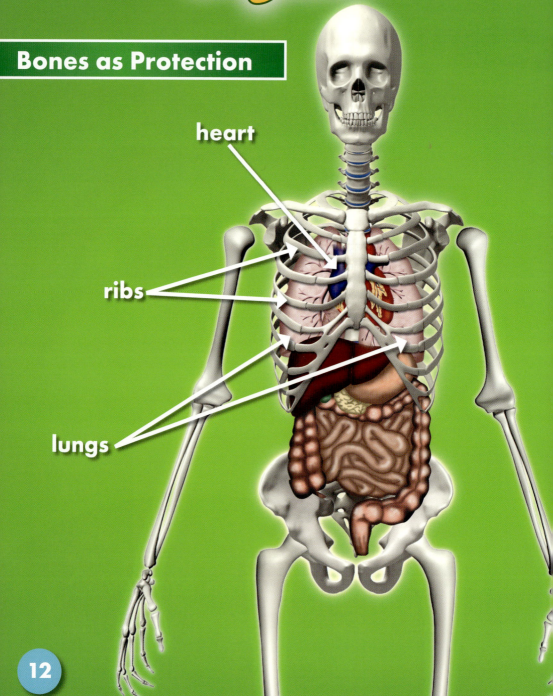

Strong bones protect softer parts of your body. The **ribs** are 12 pairs of curved bones that protect the heart and lungs.

fun fact
Bones are the hardest parts of living bodies. Bones can keep their shape long after a living thing dies.

A column of bones called **vertebrae** runs down your back. This column is called the spine. Your spine helps hold up your body. It also protects a thick bundle of nerves called the **spinal cord**. The spinal cord carries messages from your brain to your body, and from your body back up to your brain.

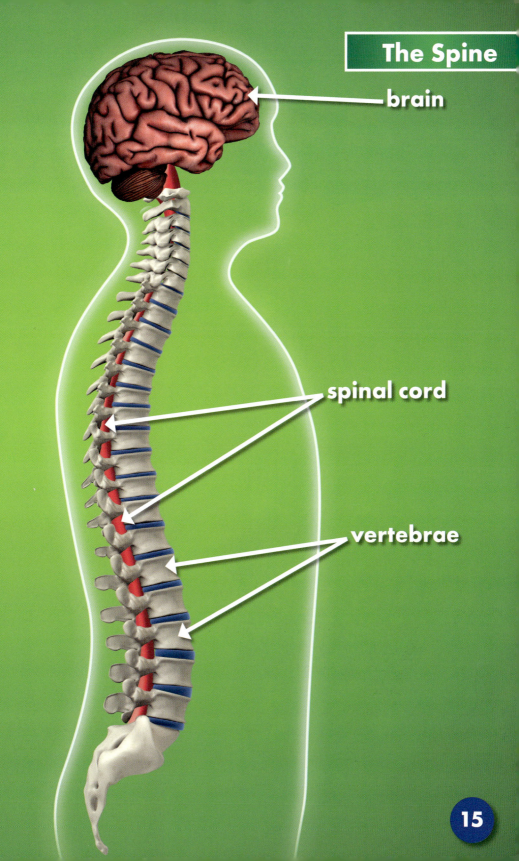

The Skull

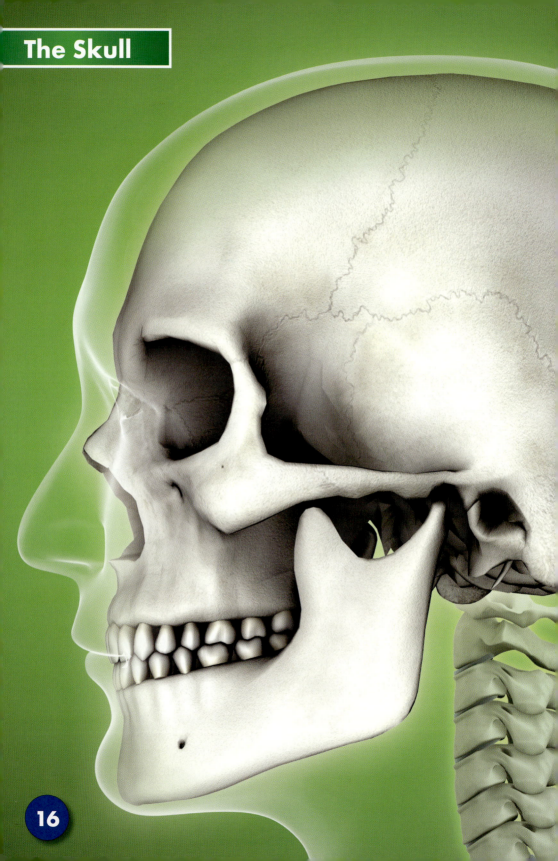

The **skull** protects your brain and gives your face its shape. The skull is made up of 29 bones. Most of them fit together tightly like pieces in a puzzle.

What's Inside Bones?

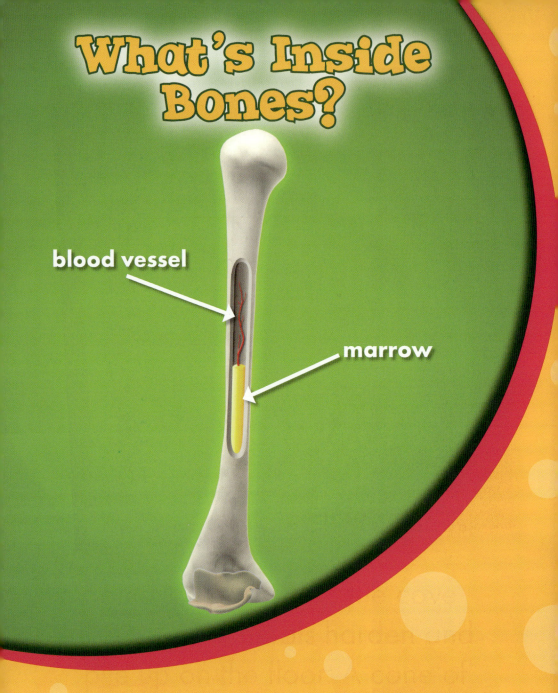

Bones also store important **nutrients** that your body needs, such as **calcium**. Your bones get nutrients from the food you eat.

Most bones also have a hollow center that holds **marrow**. This material is constantly making new **blood cells**. Each of your blood cells only lives for a short time. Your body needs a constant supply of new blood cells to stay healthy.

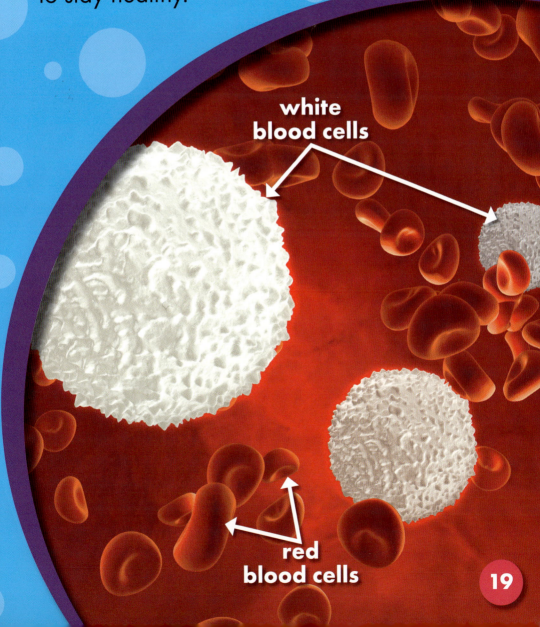

X-ray of a Broken Arm

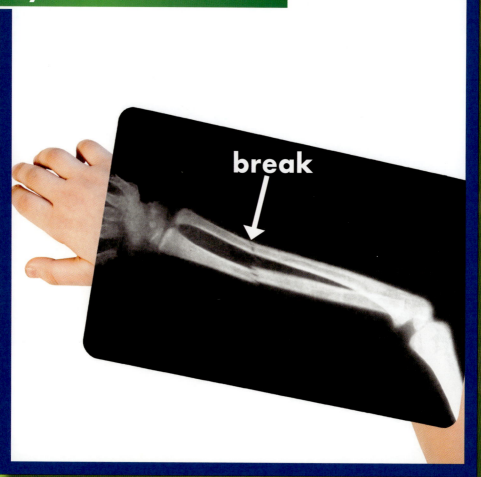

break

Like the rest of your body, bones can heal themselves when they get hurt. Doctors take an **x-ray** to see whether a bone is broken. A stiff bandage called a cast keeps the broken edges of a bone lined up correctly.

New bone grows to heal the break.
Then your bone is ready to move again!

Glossary

blood cells—small parts of blood that carry oxygen, stop bleeding, and fight sickness

calcium—a substance in some foods and in the environment that is essential for living things; calcium is stored in bones and helps to keep them strong.

cartilage—a strong but flexible tissue that provides a smooth surface along joints

joints—places where bones come together in the body

ligaments—bands of tissue that connect bones to other bones in joints

marrow—the tissue that makes new blood cells for the body; marrow is in the hollow spaces inside of bones.

muscle—a body part that can squeeze to produce force or motion

nerves—thin strings of tissue throughout the body that carry messages between the brain and other parts of the body

nutrients—substances that are necessary for living things to stay healthy

organs—parts of the body that do certain jobs to help the body function

ribs—the curved bones of the chest that protect the lungs and heart

skull—the hard, rounded set of bones in the head that protects the brain and helps shape the head and face

spinal cord—a thick cord of nerves that runs down the middle of the back

vertebrae—the small bones in the spine that protect the spinal cord

x-ray—a special photograph taken by using invisible rays that can pass through your skin; x-rays show some of the parts inside your body, such as your bones.

To Learn More

AT THE LIBRARY

Barner, Bob. *Dem Bones*. San Francisco, Calif.: Chronicle Books, 1996.

Olien, Rebecca. *The Skeletal System*. Minneapolis, Minn.: Capstone, 2006.

Simon, Seymour. *Bones: Our Skeletal System*. New York: HarperCollins, 2000.

ON THE WEB

Learning more about the skeletal system is as easy as 1, 2, 3.

1. Go to www.factsurfer.com.

2. Enter "skeletal system" into the search box.

3. Click the "Surf" button and you will see a list of related Web sites.

With factsurfer.com, finding more information is just a click away.

Index

babies, 5
blood cells, 19
body, 5, 6, 7, 9, 10, 13, 14, 18, 19, 20
brain, 11, 14, 17
calcium, 18
cartilage, 9
cast, 20
doctors, 20
face, 17
femur, 6
fingers, 7
food, 18
hands, 7
heart, 13
hinge joints, 9
joints, 9, 10
ligaments, 9
lungs, 13
marrow, 19
messages, 11, 14
movement, 5, 9, 10, 11, 21
muscles, 11

nerves, 11, 14
nutrients, 18
organs, 5
ribs, 13
shape, 5, 6, 13, 17
skull, 17
spinal cord, 14
spine, 14
stirrup, 7
thumbs, 7
vertebrae, 14
x-rays, 20

The images in this book are reproduced through the courtesy of: Sebastian Kaulitzki, front cover, pp. 4-5, 6, 7, 12, 15, 19; Monkey Business Images, pp. 4-5; Linda Bucklin, p. 8; Mandy Godbehear, pp. 10, 21; Juan Martinez, p. 11; Linda Clavel, diagram, p. 12; Anatomical Design, p. 15; Eroldemir, p. 20.